ANOTHER THIRD POETRY BOOK

compiled by John Foster

Oxford University Press

Oxford University Press, Walton Street, Oxford OX2 6DP

Oxford New York Toronto
Delhi Bombay Calcutta Madras Karachi
Kuala Lumpur Singapore Hong Kong Tokyo
Nairobi Dar es Salaam Cape Town
Melbourne Auckland Madrid

and associated companies in
Berlin Ibadan

Oxford is a trade mark of Oxford University Press

© John Foster 1988
First published in paperback 1988
Reprinted 1989, 1993, 1994

First published in hardback 1988
Reprinted 1989

ISBN 0 19 917123 8

Composition in Palatino by Pentacor plc, High Wycombe, Bucks
Printed in Hong Kong

Contents

Circus School

Roll up for the circus school!
Roll up!

The ringmaster whips for silence
And all is a hush of waiting.

Bring in the clowns!
Red-faced and loud with laughter
All in a tumble they sprawl
Late with their laces unfastened.
The cat ate their homework, sir.

A giggle of chimps
All dressed up and only school to go to
Pick and preen each other at the back.

Lovely with knowledge
The golden lions stalk and snarl
And wait for the crack of words.

Elephants
Heavy
As dictionaries
And slow
As hours
Plod to their places
Smiling and trying to please.
Clap them.

Up prance the long-legged horses
Fresh with new kicks and
Skittish with right answers
That ring as bright as harness bells.

And the girl on the trapeze
Could swing and dream all day
High above the chalk dust
In the dazzle of the light bulb.

But in bursts the juggler
Tossing his tricks
To baffle and amaze
With their glitter and precision
To make you gasp.
See! Catch! Hold!
He'll teach you all his magic
If you'll just learn how to spell.

And when the show's over
When the fanfare has faded
Far past the school gates
The caretaker comes
To sweep up the sawdust
To turn down the lights
To sigh in the silence
Before
The next
Performance
Starts.

Berlie Doherty

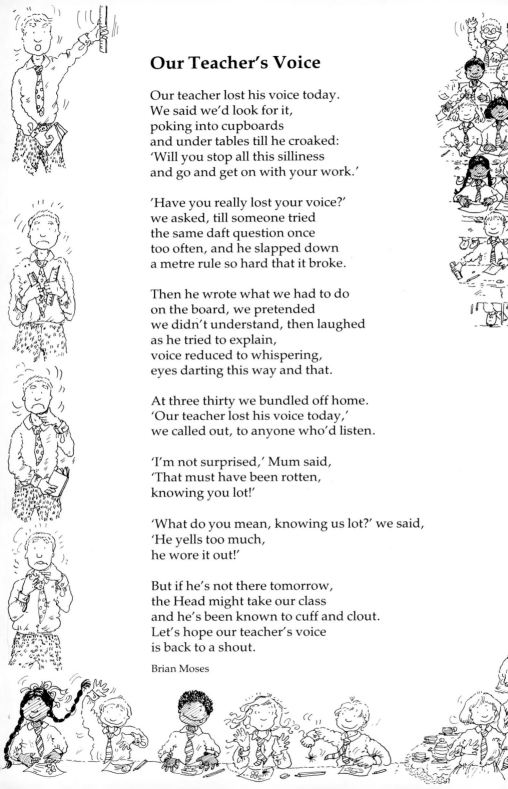

Our Teacher's Voice

Our teacher lost his voice today.
We said we'd look for it,
poking into cupboards
and under tables till he croaked:
'Will you stop all this silliness
and go and get on with your work.'

'Have you really lost your voice?'
we asked, till someone tried
the same daft question once
too often, and he slapped down
a metre rule so hard that it broke.

Then he wrote what we had to do
on the board, we pretended
we didn't understand, then laughed
as he tried to explain,
voice reduced to whispering,
eyes darting this way and that.

At three thirty we bundled off home.
'Our teacher lost his voice today,'
we called out, to anyone who'd listen.

'I'm not surprised,' Mum said,
'That must have been rotten,
knowing you lot!'

'What do you mean, knowing us lot?' we said,
'He yells too much,
he wore it out!'

But if he's not there tomorrow,
the Head might take our class
and he's been known to cuff and clout.
Let's hope our teacher's voice
is back to a shout.

Brian Moses

Question

Please tell me, dear Katy,
Why is it that in-
side the classroom you climb, jump and swing,
And somersault freely all over the floor,
The chairs and Shepperton twins.

But when we go down to
The hall for P.E.,
With ropes, horses, benches, the lot,
And masses of room to do handstands, you say,
'I'd rather just sit here and watch'.

Theresa Heine

11

Back to School Blues

Late August,
The miserable countdown starts,
Millions of kids
With lead in their hearts.
In Woolies' window: rubbers, rulers,
Geometry sets,
And a BACK TO SCHOOL sign —
I mean, who forgets?
In the clothes shops
Ghastly models of kids with
New satchels and blazers and shoes:
Enough to give anybody
Those Back to School Blues.

And Auntie Nell from Liverpool,
Who's down with us for a visit,
Smiles and says, 'So it's back to school
On Wednesday for you is it?
I only wish I'd got the chance
Of my schooldays over again . . .
Happiest days of my life they were —
Though I didn't realise it then . . . '
And she rabbits on like that,
Just twisting away at the screws;
She's forgotten about
The Back to School Blues.

And six and a half long weeks
Have melted away like ice cream:
That Costa Brava fortnight's
Vanished like a dream.
And Dad says, 'Look, this term
At school, could you try and do
A bit better?
For a start you could learn to spell
And write a decent letter.
And just keep away from that Hazel Stephens —
She's total bad news . . '
Any wonder that I've got
Those Back to School Blues?

Eric Finney

My New Goldfish

I'm taking my brand new fish to school,
Slopping about in its globe of a pool
Across the playground.

I have to say
My new fish is hardly gold,
Not quite
What I expected;
More yellow and white,
If you see what I mean.

But it's obviously been
Used to a circular house.
Quieter than any mouse
Three times clockwise round he goes,
Then a swerve,
(As if annoyed)
And three times the other way
To avoid,
I suppose,
A permanent right hand curve
From end of tail to tip of nose.

He tries to talk to me
Gulping mouthfuls of water.
I'm sure that he can see
I care
As I stand gazing in at his window
Under the map of Europe
Tossing my hair.

But the breadcrumbs at playtime were a mistake
No doubt;
He looked at me mournfully
And spat them out.

Gregory Harrison

A Hot Day at the School

All day long the sun glared
as fiercely as a cross Head teacher.

Out on the brown, parched field
we trained hard for next week's Sports Day.

Hedges wilted in the heat;
teachers' cars sweltered on the tarmac.

In the distance, a grenade of thunder
exploded across the glass sky.

Wes Magee

15

Travelling Child

He came in tabby September
When the fair shrieked on the hill,
Stood on the classroom threshold,
Stray cat on our window-sill.

His gold eyes begged for a welcome,
For milk and a place at our fire
But his thin frame remembered overarm stones,
Claws that tore like barbed-wire.

In the yard after milk, in chrysanthemum sun,
We watched, domestic as cream.
The gipsy-boy, single, stood clenched by the fence,
Dark as a midnight dream.

We did not claw, nor spit nor hiss,
Yet we never invited him in,
Closed all our windows to with a thud,
Slammed each of our doors on him.

Jacqueline Brown

New Baby

Mi baby sista come home las' week
An' little most mi dead,
When mama pull back de blanket
An' me see de pickney head.

Couple piece a hair she hab pon i',
An de little pickney face
Wrinkle up an crease up so,
It was a real disgrace.

Mi see har a chew up mama chest
So mi gi' har piece o' meat,
Mama tek i' whey, sey she cyaan eat yet
For she no hab no teeth.

Mi tell mama fi put har down
Mek she play wid mi blue van,
She sey Yvonne cyaan siddung nor stan' up yet
Nor hol' tings eena har han'.

Mi sey a' right but maybe
She can play 'I spy' wid mi,
She tell mi de pickney cyaan talk yet
An she can hardly see.

Aldoah she no hab no use,
An she always wet har bed,
Mi wouldn' mine so much ef she neva
Mek so much nize a mi head.

Every night she wake mi up;
But a mama mi sorry fah,
For everytime she wake up
She start fi eat mama.

She blind, she dumb, she ugly, she bald,
She smelly, she cyaan understan',
A wish mama would tek har back
An' buy one different one.

Valerie Bloom

18

Using a Telescope

My brother, who got a telescope for Christmas
And often explains to me,
Though I still don't understand,
What makes a star a planet,
Chalked a diagram of the universe
On a side of the house.
If the earth were a grain of sand
Pluto would be
The length of the wall away.

A wonderful thing a telescope:
We haven't seen Pluto or even Saturn
But from the bedroom window
All the family has had a turn
At reading a page of a book
Open at the bottom of the garden.

Stanley Cook

Skating on Thin Lino

Because there is no Ice Rink
Within fifty miles of our house,
My sister perfects her dance routines
In the Olympic Stadium of my bedroom.
Wearing a soft expression
And two big, yellow dusters on her feet,
She explodes out of cupboards
To an avalanche of music
And whirls about the polished lino
In a blur of double axles and triple salkoes.
For her free style doubles
She hurls this pillow called Torvill
From here to breakfast time
While spinning like a drunken hippo
Round and round my bed.
Imagine waking up to that each morning!
Small wonder my hands shake
And I'm off my cornflakes.
Last Thursday she even made me
Stand up on my bed
And hold up cards marked 'Six'
While she gave endless victory salutes
In the direction of the gerbil's cage.
To be honest,
Despite her practice and her endless dedication,
I don't think she has a hope
Of lifting the world title.
But who cares;
She may not get the gold
But I bet there isn't another skater alive
With wall to wall mirror
On her bedroom floor.

Gareth Owen

The Wrong Side

My mother used to tell me
I'd got out of bed
on the wrong side, which was strange
as there was only one side
I could tumble from.
The other was hard against the wall
and all I did was crack
my knee, but still she insisted
that she was right, so one bright
morning I tried it out, squeezed
between the wall and my bed
then said nothing. She never knew.
I was puzzled.
My mother said how she'd teach me
to choose between wrong and right,
but if I got out of the right side
and that was wrong,
then who was right?

Brian Moses

Whisper Whisper

whisper whisper
whisper whisper
goes my sister
down the phone

whisper whisper
go the beech leaves
breathing in the
wind alone

whisper whisper
whisper whisper
slips the river
on the stone

whisper whisper
go my parents
when they whisper
on their own

I don't mind the
whisper whisper
whisper whisper
it's a tune

sometimes though
I wish the whisper
whisperings would
shut up soon

Kit Wright

Once The Wind

Once the wind
said to the sea
I am sad
      And the sea said
Why
    And the wind said
Because I
am not like the sky
or like you

    So the sea said what's
so sad about that
        Lots
of things are blue
or red or other colours too
    But nothing
neither sea nor sky
can blow so strong
or sing so long as you

    And the *sea* looked sad
    So the wind said
Why

Shake Keane

The Voyage

In my mind I see once more
The horseshoe harbour,
The fishing boats at anchor,
The foam dying on the seashore.

Beyond the harbour mouth
The lean lithe schooner
Hoists her canvas higher
Readying herself to go south,

Ride the vast ocean, to fend
Off storms, plunging,
Rising, plunging, rising,
To voyage to the wide world's end.

And I, old now, am aboard
That ship, as once I
Was when I was young, eye
Bright, every hope unflawed.

Twilight falls, the schooner
Fades, dips below
The far horizon. I go
With her, and shall forever.

Albert Rowe

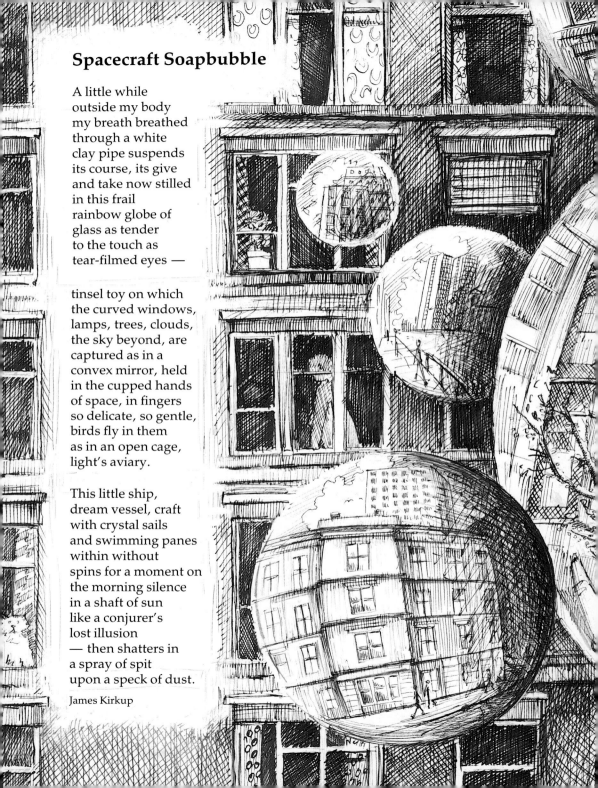

Spacecraft Soapbubble

A little while
outside my body
my breath breathed
through a white
clay pipe suspends
its course, its give
and take now stilled
in this frail
rainbow globe of
glass as tender
to the touch as
tear-filmed eyes —

tinsel toy on which
the curved windows,
lamps, trees, clouds,
the sky beyond, are
captured as in a
convex mirror, held
in the cupped hands
of space, in fingers
so delicate, so gentle,
birds fly in them
as in an open cage,
light's aviary.

This little ship,
dream vessel, craft
with crystal sails
and swimming panes
within without
spins for a moment on
the morning silence
in a shaft of sun
like a conjurer's
lost illusion
— then shatters in
a spray of spit
upon a speck of dust.

James Kirkup

The Balloonists

Hot air rises
And, their burners roaring
As they pass overhead,
The hot air balloons
Lift off from the fair,
Passengers upon the wind
Arriving they don't know when
They don't know where.

Below them
Cars keep to the road
And trains to the track
And high above them
Jetting aeroplanes
Playing noughts and crosses
With their vapour trails
Return on radar
To the runway.

Holding on to their bubble of air,
Balloonists alone
Are free to follow
Where the wind is blowing
And the clouds are going.

Stanley Cook

28

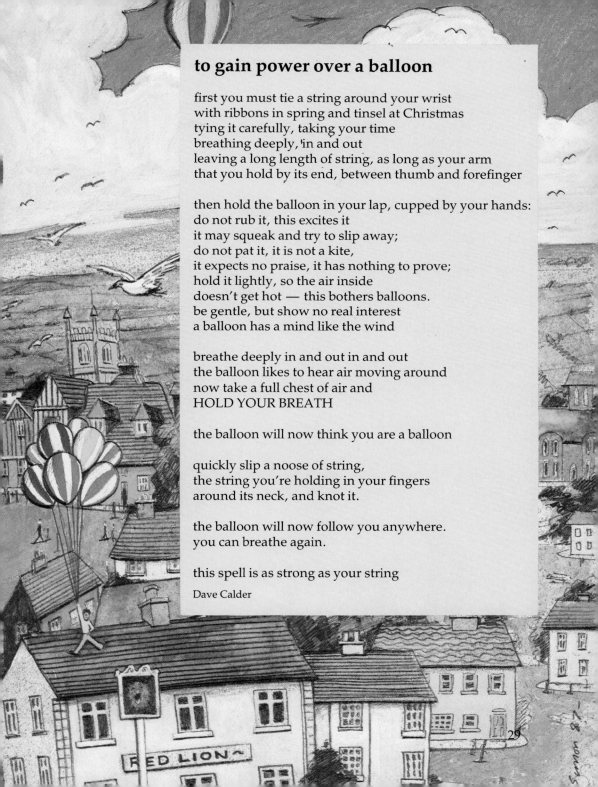

to gain power over a balloon

first you must tie a string around your wrist
with ribbons in spring and tinsel at Christmas
tying it carefully, taking your time
breathing deeply, 'in and out
leaving a long length of string, as long as your arm
that you hold by its end, between thumb and forefinger

then hold the balloon in your lap, cupped by your hands:
do not rub it, this excites it
it may squeak and try to slip away;
do not pat it, it is not a kite,
it expects no praise, it has nothing to prove;
hold it lightly, so the air inside
doesn't get hot — this bothers balloons.
be gentle, but show no real interest
a balloon has a mind like the wind

breathe deeply in and out in and out
the balloon likes to hear air moving around
now take a full chest of air and
HOLD YOUR BREATH

the balloon will now think you are a balloon

quickly slip a noose of string,
the string you're holding in your fingers
around its neck, and knot it.

the balloon will now follow you anywhere.
you can breathe again.

this spell is as strong as your string

Dave Calder

Englyn : Slowly

Slowly floating, a kite of crimson cloud —
the summer sunset light
slowly deepens into night —
slowly, a first star grows bright.

James Kirkup

What the Mountains Do

What the mountains do is
roar silent warnings over
huge brown and heather-covered spaces

or fill up valleys with dark green laughter

before resting their stone-cropped heads
in sunlight.

David Harmer

Dandelion

wish in the wind
grey globe
blow round the clock
till you burn gold again

Richard Burns

What's the time, Mr. Dandelion?

Time to spread yellow suns
over the fields,

Time to blow minute-stars
into the sky;

Time to lie counting
lions'-teeth hours,

Time to stand still now,
watching time fly.

Judith Nicholls

31

Windy Garden

The cherry's a swinging carnival.

Pyjamas wild as wrestling boys
sprout each other's limbs.

A green nightdress
stands on its invisible head.

There's a swallow
shadow scything.

The ghost of a drunk sailor
on the treehouse ladder.

Watched by incredulous crows —

Two crows.

Geoffrey Holloway

33

Haikus

The larch cone, an owl
Sitting on a branch with feathers
Ruffled, fast asleep

Juliet Mayo

Larch — a Christmas tree
Decorations green as grass
With roses, lucky me

Gary Wade

An acorn has fallen
From a very old oak, leaving
A small green pipe

Colin Bales

See these hazelnuts,
Hanging, like a group of bells,
Ringing in the wind.

James Langton

A tiny helicopter
Flying in the massed green clouds
Of the sycamore tree.

Stuart Holding

Wings of a butterfly
Flying over the garden on
An autumn day

Marianne Frall

Telephone Wires

Jacqueline Brown

Autumn Sorrow

The moon sobs
    open-mouthed
and stars hang
    like tears
on the cheeks of night.

Moira Andrew

Haiku

A bitter morning:
    sparrows sitting together
      without any necks.

J.W. Hackett

Autumn Horseman

He comes on his piebald
Spotted yellow and brown.
He dismounts, sits down
Sighing among us. He

Rustles as he moves,
Coat in tatters, sudden
Crack of knees, like sodden
Sticks reluctant to burn.

He coughs, speaks grave words.
Dry leaves drifting are shed
From his bent frosted head.
He mentions clouds and storms.

Our minds darken, become
Cold, quiet, withdrawn,
Counting each icy dawn
We'll go numb into the fields.

He smiles, leaves us fruits,
Sun-ripened grain.
He says, 'There's no gain
Without loss.' He rises,

As the wakened wind rises,
Shivers, mounts his pony.
We gather his gifts gratefully.
We feel like weeping.

Albert Rowe

Umbrella Poems

Spring Rain

Ring of drops
on drum of oiled
paper, notes
tapped by time's
warning fingers.

Summer Gale

Before me like a shield
behind me like a sail —
what was to have been
a sunshade.

Autumn Leaves

Fall, stroking my paper roof
with crimson sighs, or lonely hands
lighter than gingko fans,
screening my golden day
with a moment's dark designs.

Winter Snow

Fast falling flakes —
I open my umbrella
only half-way:
steep snow-country roof.

James Kirkup

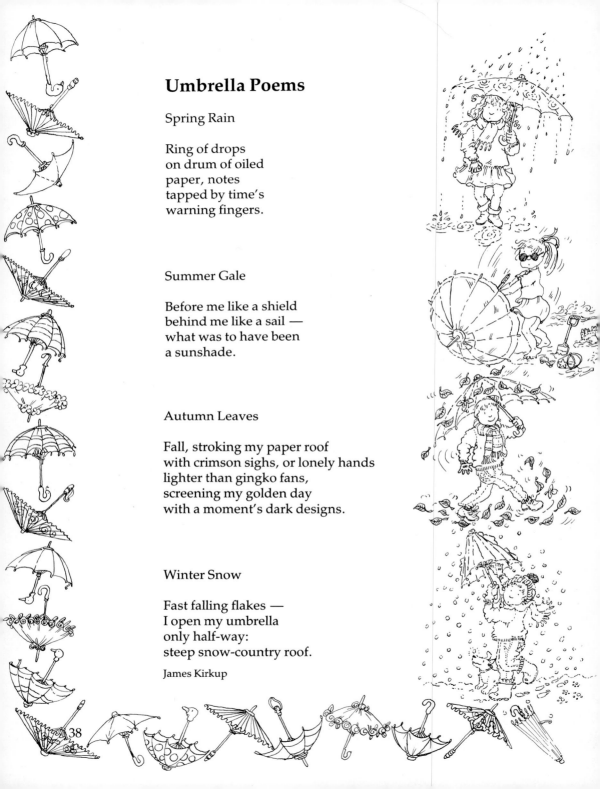

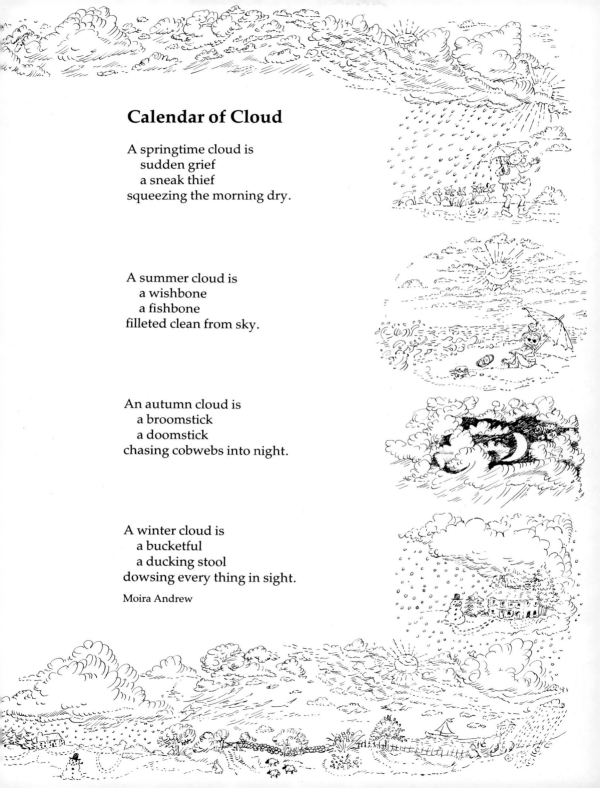

Calendar of Cloud

A springtime cloud is
   sudden grief
   a sneak thief
squeezing the morning dry.

A summer cloud is
   a wishbone
   a fishbone
filleted clean from sky.

An autumn cloud is
   a broomstick
   a doomstick
chasing cobwebs into night.

A winter cloud is
   a bucketful
   a ducking stool
dowsing every thing in sight.

Moira Andrew

Winter Journey

Fog gathers over rivers like steam on cauldrons
And catches in hawthorns and hemlock skeletons,
Torn off a single enormous cobweb.
The leafless trees start up like highwaymen
Beside the road and for a week
Above the same altitude
Visibility has almost ceased.
Smoke is lost and lights are faded out
In the vague cottages and looming farms;
Not a soul is about
As if, at the onset of a new Ice Age,
They had all gone south.

Stanley Cook

January Moon

The moon skates through the sky
Against the wind, against
The low-flying clouds.
On blue ice she dances
Soundlessly.

Gerda Mayer

Simple Seasons

Swallows,
Primroses
Return.
It's
New,
Green!

Skylarks
Up,
Meadows
Motley,
Elms
Regal.

Apples
Untold,
Trees
Unruly;
Mists
Now.

Waters
Icebound,
Naked
Trees;
Earth
Rests.

Eric Finney

Magpie in the Snow

White land
Black veins of branches
Dead blue eye of the sky
Magpie flicks tail
Dances
Winks a living eye.

East wind
Dry bones of branches
Scoured and aching sky
Magpie cocks head
Listens
Views the world awry.

Hard ground
Thin roof of branches
Far unfriendly sky
Magpie cares naught
Chatters
Flings its wings to fly.

Michael Tanner

Magpie

Magpie.
Bandit,
Callous as the carrion crow
Decimator of ducklings
Innocent as thistledown.

Magpie.
Boss bird,
You brought off a take-over
Bid for our peaceful garden:
Piebald villain, on guard still.

Magpie,
No frail
Egg or soft nestling is safe —
Wren, robin, song thrush, blackbird —
From your hungry eye and beak.

Magpie.
Then, though
Wary still, you allow me
Close, as if you know I wish
To contemplate the beauty,

Magpie,
Often
I wonder why I welcome
Your spick-and-span swagger,
Flirting your long perky tail,

Magpie,
Shyly
Hidden, of breast, pinion, tail
Iridescent in the sun,
As new-minted there you stand.

Albert Rowe

Magpie,
Hating
Your crimes. It's that you are brave.
With your mate you see off each
Rival predatory crow,

Explorer

Two o'clock:
Let out of the back door of the house, our cat
Is practising the snow.

The layer of white makes a small straight, crumbling cliff
Where we open the back door inwards. The cat
Sniffs at it with suspicion, learns you can just about
Pat at the flaking snow with a careful dab. Then,
A little bolder, he dints it with one whole foot
— And withdraws it, curls it as if slightly lame,

And looks down at it, oddly. The snow is
Different from anything else, not like
A rug, or a stretch of lino, or an armchair to claw upon
And be told to *Get off!*

The snow is peculiar, but not forbidden. The cat
Is welcome to go out in the snow. Does
The snow welcome the cat?
He thinks, looks, tries again.

Three paces out of the door, his white feet find
You sink a little way all of the time, it is slow and cold,
    but it
Doesn't particularly hurt. Perhaps you can even enjoy
    it as something new.
So he walks on, precisely, on the tips of very
    cautious paws . . .

Half past three, the cat stretched warm indoors,
From the bedroom window we can see his explorations

— From door to fence, from fence to gate, from gate to
    wall to tree, and back,
Are long patterned tracks and trade-routes of round
    paw-marks
Which fresh snow is quietly filling.

Alan Brownjohn

Under Frost

Under frost
And snow
Still earth
And seed
Dark growth
Green hint
Of Spring
And all
That Spring
Can bring.

John Kitching

46

The Signal

It catches your eye,
that sudden flash of sunlight
from a window
in the distant block of flats.

Again, the signal,
a seasonal message
to say that Spring has arrived,
winter gone.

Now the air warms,
birds yammer ceaselessly.
Blades of grass and leaves
yawn, stretch.

Perhaps once a year
it catches your eye,
that sudden flash of sunlight
from a distant window.

Wes Magee

47

Green Man in the Garden

Green man in the garden
    Staring from the tree,
Why do you look so long and hard
    Through the pane at me?

Your eyes are dark as holly,
    Of sycamore your horns,
Your bones are made of elder-branch,
    Your teeth are made of thorns.

Your hat is made of ivy-leaf,
    Of bark your dancing shoes,
And evergreen and green and green
    Your jacket and shirt and trews.

Leave your house and leave your land
    And throw away the key,
And never look behind, he creaked,
    And come and live with me.

I bolted up the window,
    I bolted up the door,
I drew the blind that I should find
    The green man never more.

But when I softly turned the stair
    As I went up to bed,
I saw the green man standing there.
    'Sleep well, my friend,' he said.

Charles Causley

Mystery

Was it really?
You never did!
At the top of the house?
But where was Sid?
How big were the footmarks?
What kind of a noise?
What on earth do you mean —
With the children's toys?
The *whole* of the sofa?
And the easy chairs?
I can't believe it!
Under the stairs?
Never the frying pan!
And the dustbin lid?
What I don't understand is
Where was Sid?

John Kitching

Uncle Fred

'There's no such things as ghosts' —
That's what my uncle Fred said
'There's no such things as ghosts
no ghastly ghouls
or ghostly spooks
no wizened witches
nor weird werewolfs
and those that are buried stay buried,'
said my Uncle Fred
'and things that go bump in the night
have fallen off the mantelpiece because the cat
knocked them over
all those pale apparitions
are just superstitions
There's no such things as ghosts.'

But then Fred died
Poor Uncle Fred
Dead
He died with a smile on his face
I went to his funeral to say goodbye
And when that night I climbed into bed
I heard a small voice
quite close to my head
'There's no such things as ghosts'
was all that it said.

Robert Fisher

A Final Appointment

Enter the servant Abdul
His face ashy grey,
Fear in his eyes —
He has seen Death today.

Begs release of his master,
Plans instant flight:
'I must be far from
This city tonight!'

'Why?' asks the Sultan,
A man kind and clever,
'You have said many times
You would serve me for ever.'

'Master, I love you,
That much you must know,
But down in the city
A half hour ago

Death himself was out walking,
Reached cold hands to me:
The threat was quite plain
For the whole world to see.

I must leave Death behind!
To Baghdad I'll take flight.
Master, give me a horse —
I can be there tonight!'

So Abdul escapes,
Fear driving him on,
And very soon after
His servant has gone

The Sultan himself
Walks out in the city,
Walks among cripples
And beggars with pity;

Like Abdul, meets Death
As he walks in that place,
Peers into the folds of his cloak
For his face;

Sees it not; hears a voice
That is cold, clear and dry:
'Look not for my face —
See that and you die.'

But the Sultan speaks boldly
Asking Death, 'For what cause
Did you threaten this morning
To make Abdul yours?'

Death replied, 'To your servant
I issued no threat.
Indeed sir, I knew that
His time was not yet.

This morning your servant
Had nothing to fear;
I was taken aback
To see the man here;

Gave a start of surprise,
Knowing well that I had
An appointment with Abdul
Tonight in Baghdad.'

Eric Finney

Wilkins' Drop

Great news!
To fulfil a lifelong ambition
Wilkins jumps from a plane;
Makes a slight omission
That's not such good news,
In fact, a proper brute:
What's missing is
His parachute.
Better news, though,
As earth comes rushing:
Haystack below —
Possible cushion.
Bad news again:
Embedded in stack —
Pitchfork with prongs
As sharp as a tack.
Good news:
Misses pitchfork. Phew!
Bad news:
Misses haystack too.

Eric Finney

Wilkins' Luck

Not liking work much,
Wilkins looked instead
For bubble schemes
To make his bread.
Heard of an island
Where reside
A peaceful people —
All one-eyed!
Planned thus: sail there,
Seize a few;
These Cyclopean freaks
On view
Back home
(At a stiff admission fee)
Would make his fortune
Easily.
Wilkins took ship
And after many a while
Arrived at the
Green Pacific isle.
Discovered what he'd heard
Was true:
Each person had a single eye -
And it was blue.
While Wilkins studied
This strange feature:
'Look!' cried the islanders,
'A two-eyed creature!'
One smarter than the rest
Was heard to say,
'To see a freak like that . .
Would people pay?'
Now Wilkins, caged,
Is twice a day on view,
And wishes that he'd thought
The matter through.

Eric Finney

Lost in a Crowd

I wanted
to be different.
Not ordinary
but famous,
so people would notice.

My friend consulted
his spell book
on
how to grow two heads
and
how to breathe fire.
What to take to
bring on warts.
Correct projection
of blood curdling screams
without the risk
of laryngitis.

Finally
he decided
to make me invisible
but people
still don't notice me.

John C. Desmond

58

The Darkling Elves

In wildest woods, on treetop shelves,
sit evil beings with evil selves—
they are the dreaded darkling elves
and they are always hungry.

In garish garb of capes and hoods,
they wait and watch within their woods
to peel your flesh and steal your goods
for they are always hungry.

Through brightest days and darkest nights
these terrifying tiny sprites
await to strike and take their bites
for they are always hungry.

Watch every leaf of every tree,
for once they pounce you cannot flee—
their teeth are sharp as sharp can be . . .
and they are always hungry.

Jack Prelutsky

The Invisible Beast

The beast that is invisible
is stalking through the park,
but you cannot see it coming
though it isn't very dark.
Oh you know it's out there somewhere
though just why you cannot tell,
but although you cannot see it
it can see you very well.

You sense its frightful features
and its great ungainly form,
and you wish that you were home now
where it's cozy, safe and warm.
And you know it's coming closer
for you smell its awful smell,
and although you cannot see it
it can see you very well.

Oh your heart is beating faster,
beating louder than a drum,
for you hear its footsteps falling
and your body's frozen numb.
And you cannot scream for terror
and your fear you cannot quell,
for although you cannot see it
it can see you very well.

Jack Prelutsky

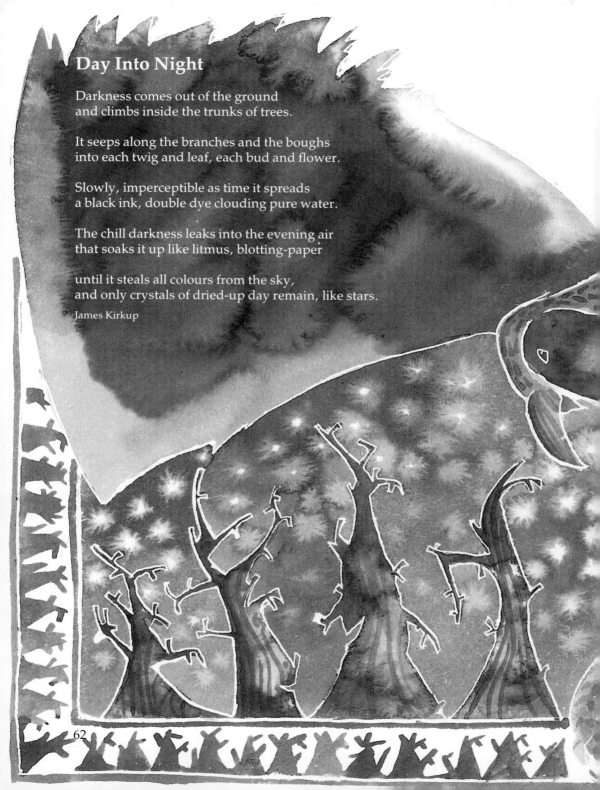

Day Into Night

Darkness comes out of the ground
and climbs inside the trunks of trees.

It seeps along the branches and the boughs
into each twig and leaf, each bud and flower.

Slowly, imperceptible as time it spreads
a black ink, double dye clouding pure water.

The chill darkness leaks into the evening air
that soaks it up like litmus, blotting-paper

until it steals all colours from the sky,
and only crystals of dried-up day remain, like stars.

James Kirkup

Darkness

Children, we burrowed beneath blankets
And knew the bed's hot darkness before
Sheets entwined our heads, left us gasping.

Worse still was waking after midnight
To glimpse a gaunt man behind the door
Where my dressing-gown usually hung.

Once, a girl and I crawled into the
Tar blackness of an air-raid shelter
Then fled as a feathered-thing squawked, flew.

Now *my* children creep down that same path,
Test darkness, tiptoe the line between
Inquisitiveness and speechless fear.

While touring old coastal defences
They insist on descending steep steps
To explore a dank, brick-strewn tunnel.

I demur, preferring the seascape
With its total sound-surround of gulls
But am dragged to a mildewy gloom

Where voices echo and birds have died.
Again that icy shiver. Silenced
We cling to each other, stumble deep

Into a tomb. Wide eyes conjure up
Unspeakable shapes in the foul air
And a finger of bone, beckoning.

Wes Magee

The House at Night

Some stealthy spider is weaving round my bed
and mice are nibbling the curtains overhead.
Weird footsteps make the floorboards crack,
the staircase creaks, chill draughts thrill down my back
from some forgotten window out of sight —
    this is the house at night.

There's a whispering on the landing
where a creepy tropic plant is standing,
and the coatrack in the hall
lets fall a scarf — a long, soft fall:
a snake's loose coils that rapidly grow tight —
    this is the house at night.

From the distant kitchen come the notes
of dripping taps, plink-plonking secret codes
I cannot get the meaning of: a sudden
icy shudder — the refrigerator groans — a hidden
oven, cooling, ticks in rustling ember-light —
    this is the house at night.

— But even stranger is my own tense breathing
as I lie here speechless looking at the ceiling
that seems to swim all round like falling snow.
I can hear my eyelids batting gently, slow —
then quick as heartbeats as I freeze with fright
at something in the mirror shining bright —
has someone left the telly on all night?
No, thank heaven, it's all right,
it's only the moon's pale, spooky light
touching my tangled sheets with chalky white —
    yes, this is the house at night.

James Kirkup

The Haunted House

The moon hobbles across the moors
like an old man heaving
a mill-stone through the sky.

Battered doors, the windows smashed,
the house shakes as thunder thumps
heavy fists on its head.

Walk nearer if you dare,
open the gate that hangs from a nail
push through the cobwebbed leaves.

Tread softly up the pathway slabbed
with gravestones, read the names,
visitors who never left.

The door creaks, footsteps clatter
down an empty corridor,
laughter gurgles up the stairs.

Run through the rooms
chased by the old man with his sack
feel long fingers at your throat.

Crawl into the attic, dust chokes
and clogs your throat. The trapdoor slams.
The house snaps tight its broken teeth

gulps you down inside.

David Harmer

Owl

Owl
was darker
than ebony —
flew through the night
eyes like amber searchlights,
rested on a post,
feathers wind-ruffled,
stood stump still,
talons ready to seize
and squeeze.

Owl
was death
for it flew through the dark
that swamped the fields,
that tightened its knot,
that bandaged the hills
in a blindfold of fear.

Owl flew — Who — Who — Who —

Pie Corbett

The Snow-shoe Hare

The Snow-Shoe Hare
in his own sudden blizzard.

Or he comes, limping after the snowstorm,
A big, lost, left-behind snowflake
Crippled with bandages.

White, he is looking for a great whiteness
To hide in.
But the starry night is on his track –

His own dogged shadow
Panics him to right, and to left, and backwards,
    and forwards –
Till he skids skittering
Out over the blue ice, meeting the Moon.

He stretches, craning slender
Listening
For the Fox's icicles and the White Owl's slow cloud.

In his popping eyes
The whole crowded heaven struggles softly.

Glassy mountains, breathless, brittle forests
Are frosty aerials
Balanced in his ears.

And his nose bobs wilder
And his hot red heart thuds harder

Tethered so tightly
To his crouching shadow.

Ted Hughes

Silver Grey

a silver grey ripple
on
a grey silver river:
the squirrel crosses the road

John Rice

Grey Squirrel

Noses against the classroom windows,
teacher standing behind us, we stare out
as a grey squirrel nimbles its way
over the field's million sodden leaves
on this damp November day.

The trees drip, the grass is dank,
the playground shines like plastic.
A bedraggled sun. All's still, still,
except for that squirrel now busy
at husks of beech nuts, nibbling his fill.

Suddenly he's bolt upright, sniffing,
and then gone, swarming up a tree trunk
like Spiderman scaling a vertical wall.
Now he tight-rope-runs along a branch
and leaps to the next tree, does not fall.

We return to our tables. Chairs scrape.
Teacher stands at the board, chalk poised.
No one speaks. For a minute we secretly gloat
over the wonder of that squirrel
in leather gloves and grey fur coat.

Wes Magee

The Field of the Rose

in the field with bushes
of the wild dog-rose
he crouches low in his form

crouches there
till the air is clear
then away he goes
at a gentle trot
lipperty-lop, lipperty-lop
at a gentle trot he goes

his movement is easy
his ears stand tall
but his nose is aflare
and his eyes survey all
as over the field
at a gentle trot
lipperty-lop, lipperty-lop
over the field
at a gentle trot he goes

at the first hint of danger
he drops to the ground
drops to a stone-like pose

then should threat
become pressing
his ears lie flat
he tenses to thrust
his legs hit back
sensing survival
he stretches
he pounds
at the earth
with the drum
and the ground
echoes loud
with the drum
of his heart
with the beat
of his pulse
with the surging
of blood
with the thundering air
in the flight
for his life
through the field
through the field
with the wild dog-rose

and the rose becomes flushed
in the evening sun

the field of the rose
is stained red

Joan Poulson

A Tomcat Is

Nightwatchman of corners
Caretaker of naps
Leg-wrestler of pillows
Depresser of laps

A master at whining
And dining on mouse
Afraid of the shadows
That hide in the house

The bird-watching bandit
On needle-point claws
The chief of detectives
On marshmallow paws

A crafty yarn-spinner
A stringer high-strung
A buttermilk moustache
A sandpaper tongue

The dude in the alley
The duke on the couch
Affectionate fellow
Occasional grouch

J. Patrick Lewis

Night Walk

What are you doing away up there
On your great long legs in the lonely air?
    Come down here, where the scents are sweet,
    Swirling around your great, wide feet.

How can you know of the urgent grass
And the whiff of the wind that will whisper and pass
    Or the lure of the dark of the garden hedge
    Or the trail of a cat on the road's black edge?

What are you doing away up there
On your great long legs in the lonely air?
    You miss so much at your great, great height
    When the ground is full of the smells of night.

Hurry then, quickly, and slacken my lead
For the mysteries speak and the messages speed
    With the talking stick and the stone's slow mirth
    That four feet find on the secret earth.

Max Fatchen

A Short Cut . . . . . After Dark

It's late.
The night is icy
as we head home
after carol singing,
coins chinking in a tin.
It's so cold;
fingers frost-bitten,
the estate quiet,
no one about.
Snow lies thinly
on pavements.
Far off a dog barks.

It's late.
We take a short cut
through the unlit
school grounds.
Climb the wall, and drop.
Race past the
'No Trespassers' sign,
the skeletal trees,
the bushes hunched
like sleeping bears.
Beneath our Wellingtons
snow creaks.

Hearts thump as,
breathless, we stop
at the school building,
its windows dark.
Inhale. Your lungs hurt
with freezing air.
Listen. An owl hoots.
In the clear sky
a million stars
are like silver nails
hammered into the hull
of a vast, black ship.

The last lap.
Wraith-like we skate
over the playground,
vault the padlocked gate.
Back on the street.
No cars. No people.
Three days to Christmas
and our carols long gone
into the frozen night.
Lights in the hall.
Home. Warmth.
It's late.

Wes Magee

Whoppers

'I'm having a pony for Christmas,
  And a meal at a posh hotel.'
'That's nothing, I'm having video,
  And two colour tellies as well.'

'My dad's having a Rolls Royce car.'
'Well, my dad's having two —
  One for his window-cleaning gear
  And one for mum — brand new.'

'My mum's having a baby.'
'Well, my mum's having twins —
  Or maybe she'll have triplets,
  Or even quads or quins.'

'I'm having a sailing dinghy:
  Cor, won't the neighbours go green!'
'We're having the yacht Britannia
  Bought secondhand from the Queen.'

'We're off to the Costa Brava,
  Dad's getting tickets quite soon.'
'I'll think of you then while we're on
  Our luxury tour of the moon.'

. . . . .

'To tell you the truth, I've been fibbing
  And boasting, I realize.'
'That's nothing: I've not been telling fibs,
  But monstrous, walloping lies!'

Eric Finney

78

Lies

When we are bored
My friend and I
Tell
Lies.

It's a competition: the prize
Is won by the one
Whose lies
Are the bigger size.

We really do:
That's true
But there isn't a prize:
That's lies.

Kit Wright

Christmus a Come (Christmas is Coming)

De fee-fee dem come out big an bright,
An everywhey yuh go
Pickney a pick de fee-fee dem
An a blow dem loud yuh know.

De poinsettia leaf dem change from green
An tun as red as blood,
One turkey eena de fowl coop
A get fatten up fe food.

De sorrel red, it soon ready
Fe bwile up wid de wine,
An Papa sey de gungo peas
An yellow yam doing fine.

So we soon start fe have gungo soup.
De orange dem a ripe,
An breeze start blow soh cold now
Mi haffe wear sweater go a stan-pipe.

Me sista a mek Gran-market frock
Mama soaking fruits fe bake
An Pape buy one crate a wine
An some rum fe soak de cake.

De ginger dem a blossom,
Me eat one tangerine today,
An only two week lef now
Before school gi holiday.

Everytime me turn on de radio
Me hear Chrismus carol a play,
So me know fe sure sey Chrismus a come,
Yes, bwoy, Chrismus pon de way.

Valerie Bloom

My New Year's Resolutions

I will not throw the cat out the window
Or put a frog in my sister's bed
I will not tie my brother's shoelaces together
Nor jump from the roof of Dad's shed.
I shall remember my aunt's next birthday
And tidy my room once a week
I'll not moan at Mum's cooking (Ugh! fish fingers again!)
Nor give her any more of my cheek.
I will not pick my nose if I can help it
I shall fold up my clothes, comb my hair,
I will say please and thank you (even when I don't mean it)
And never spit or shout or even swear.
I shall write each day in my diary
Try my hardest to be helpful at school
I shall help old ladies cross the roads (even if they don't want to)
And when others are rude I'll stay cool.
I'll go to bed with the owls and be up with the larks
And close every door behind me
I shall squeeze from the bottom of every toothpaste tube
And stay where trouble can't find me.
I shall start again, turn over a new leaf,
Leave my bad old ways forever
Shall I start them this year, or next year
Shall I sometime, or . . . . . ?

Robert Fisher

So Clever

Computers are remarkable,
Say electronic highbrows
But none is clever as the one
That's right above your eyebrows.

So Help Me!

Computers are programmed
With never a blister.
I wish a computer
Would programme my sister.

Max Fatchen

What a Calamity!

Little Harold, I'll be frank,
Fell in a computer bank.
No-one knows how it occurred.
Operators there conferred.
No-one laughed or even smiled.
WHERE was little Harold filed?

It might take a day or week
Electronic hide-and-seek,
Keeping this poor boy in mind,
Pressing buttons 'Search' and 'Find.'

Then a friendly green light glowed,
For, at last, they'd found the code.
A sudden clatter, then a shout
And there was Harold . . . printed out.

Max Fatchen

Know-All

They built — it had to come one day —
A vast computer,
Fed it from A
To Z all human knowledge,
Nothing was omitted;
Then gathered round
All kinds of men from wise
To muddle-witted.
They asked it one question:
(All were in accord,
Had had it in their minds
Throughout)
'Is there a God?'
From the machine
No human furrowing of brow,
Simply in microseconds
Upon the screen
The answer
'There is now.'

Eric Finney

Seven Riddles

I know just how to make you fly:
I poke a fat thumb in one eye,
A finger in the other one —
It's not, for you, a load of fun;
But off you go now, grinding, gnashing,
Your ballet dancer's legs a-flashing.

Tapper, swisher, prodder, prop:
Never walks, can only hop.

Pinky brown, then farther in
Green, then white and skin on skin.
So to the knife, the pot is calling —
It's not for sadness tears are falling.

It's after the asking,
It's nearly a noose,
It's a crook,
It's a hook
With a blob that's come loose.

Like a tadpole here you sit,
Saying, 'Wait here just a bit.'

Spine like a man,
Leaves like a tree,
Jacket like a hunstman,
But is none of these three.

Poor, pale fellow:
You'll run
In the sun.

Eric Finney

Punishment

An orange instead of an egg —
That's what the brown hen made!
And the chick's astonished comment was:
'Look what marmalade!'

Got this smashing new
Continental quilt
In stripes of pink and grey,
And now sheets, blankets,
Eiderdowns
Don't seem necessary
Duvet?

You may think I'm
Joking
When I tell you
I'm buying
'Lord of the Rings'
With my
Book Tolkien.

We've got this sauce competition
Going in our family:
I'm half-way down the salad cream,
Dad's nearly finished his H.P.
Mum, though, has only just started
Her favourite sauce:
It's up tomato ketchup, of course.

Eric Finney

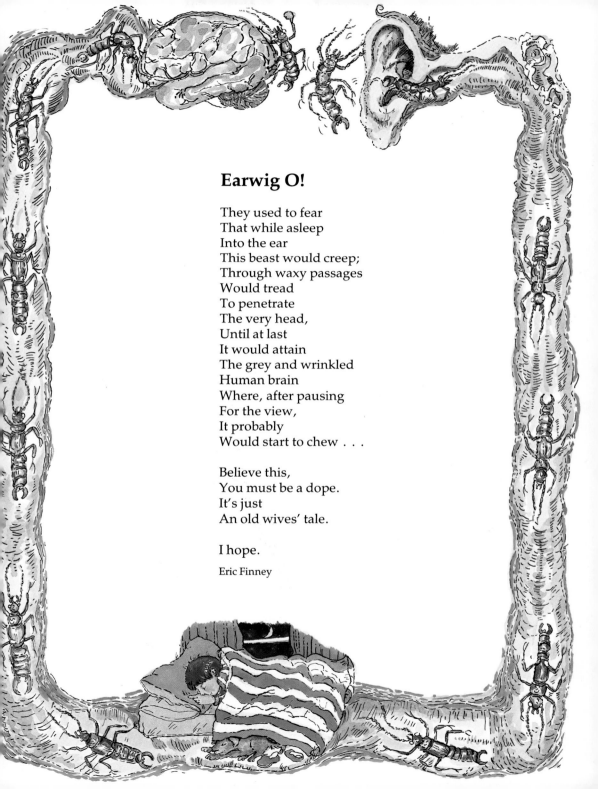

Earwig O!

They used to fear
That while asleep
Into the ear
This beast would creep;
Through waxy passages
Would tread
To penetrate
The very head,
Until at last
It would attain
The grey and wrinkled
Human brain
Where, after pausing
For the view,
It probably
Would start to chew . . .

Believe this,
You must be a dope.
It's just
An old wives' tale.

I hope.

Eric Finney

The Chimpanzee

The chimpanzee
can climb a tree
in a highly professional fashion
and scoff bananas for his tea
with a love that amounts to a passion.

The chimpzeepan
does much less than he can
and he drives nearly everyone crazy
and he hasn't been seen to move for years.
Is he dead, or just dreadfully lazy?

The zeepanchimp
has a bit of a limp
which makes him considerably slower,
and he hasn't a very good head for heights
so he stays on the ground, cos it's lower.

John Bond

The Ostrich

The ostrich roams the great Sahara.
Its mouth is wide, its neck is narra.
It has such long and lofty legs,
I'm glad it sits to lay its eggs.

Ogden Nash

Wolf

I've seen a chimp
wolf bananas,
but I've never seen a wolf
chimp anything.

John Bond

Demolition Worker

There he is, ten storeys above the street,
Highlighted by his white shirtsleeves,
No hard hat or safety harness
But pick in hand, on top of a narrow peak
Made, not of stone, but of brick.
His way to get from the roof to the ground
Is to knock the building beneath him down,
Like knocking a mountain bit by bit
From underneath your feet
As a means of descending it.
From the way he walks on the wall, pausing to kick
Mortar down with his steel-toed boot,
Everest would seem easy to him.

Stanley Cook

It's a Fact!

The highest land creature
ever recorded
was a jumping spider
found on Everest
at 22 000 feet.

I wish that I
could jump that high!

John Foster

Mosquito

A mosquito
from Buckingham Palace
to its Maker
claimed to have been:
By Special Appointment Purveyor
of Mosquito Bites
To the Queen.

N.M. Bodecker

Look — Said the Boy

Look — said the boy
the scaffold- man at work
is like a spider on his net

No — said the scaffold-man
I'm just a fly
in the trap the spider set

Michael Rosen

Adventure

The moon said:
'I will show you gardens more lavish than the sun's;
Flowers more magical;
Stranger enchantments; heavier odours.
Come.'

And the butterfly followed
Down to a distant sea;
And perished
Trying to perch on the foaming blossoms
Of moonlit waves.

Louis Untermeyer

The Silkworms' Undoing

From moon to noon
And noon to moon
And moon to noon
The silkworms twist
And turn and spin
To spin a silk cocoon
To change a crawl for wings within
To change a shape and skin within
To turn to turn to turn within
To Moth.
They steal away
That womb that womb
They turn it to
A tomb a tomb
They unwind
The silk cocoon
And turn the silk
On loom and loom
On loom and loom
They turn the silk
And turn the silk
To cloth.

Julie Holder

95

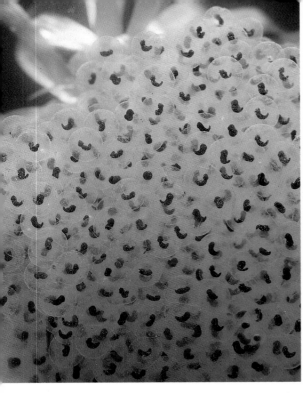

Frogspawn, I love it

Frogspawn, I love it.
It's like sago pudding in washing-up water.
I can't resist the stuff.
Jelly, gone grey.
I ladle it into my jar
By the fistful
So it oozes between my fingers.
That's how much I love it.
Then I walk home with it wobbling
Swaying slime
Like a tide, seething with secrets.
I dollop it into our pond
And there in black water it
Spreads and sinks and settles
Glutinous as a parachute.
And I wait for the bubbles to burst.

Out they come —
Tadpoles in a swarm
A bobbing of baby whales
Demi-semi-quavers all in a dash
Blobs on the boil.
Dots though they are, they bud arms
And kick their tails away with sudden legs.
They're not black any more, or brown,
Like shiny sweets to suck
They're tawny-brown, they're yellow-green,
They're the colour of rivers, and lettuces,
and onion tops.

They're frogs.
Belly-bulging,
Swell-chinned,
Kangaroo-legged,
Pop-eyed.
They flop onto rocks
And blink as the world swims past.
They dive-bomb down into darkness
And lurk in caves.
They sing like balloons.

But soon
Tiny as thumbnails
They launch themselves out
Out into the tall grasses
Where monster cats stalk
And giant birds hover
And lawnmowers prowl.

I count them as they go
And count them home, and know
However long I wait for them
Some little frogs
Will never come.

Berlie Doherty

Frogspawn

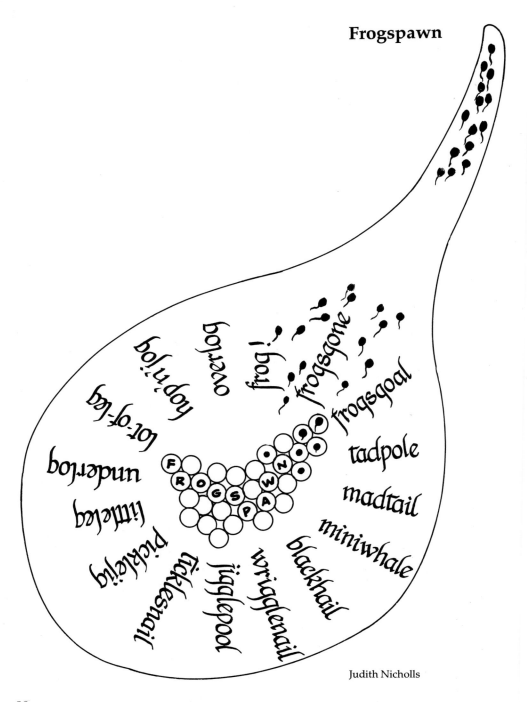

hoppriog

overlog

frog!

frogsgone

frogsgoal

lot-of-leg

tadpole

underlog

madtail

littleleg

miniwhale

prickleg

blacktail

ticklesnail

wrigglenail

jigglepool

FROGSPAWN

Judith Nicholls

98

Frog Fable

Two frogs fell, splash
into a bucket of cream. The first
(a pessimist)
bibble-babble-bubble sank to the bottom
and drowned. The other
took more trouble than his brother.
With thrusting legs he thrashed out
into the clotting cream. 'I'll survive!'
he vowed. And to his purpose sticking,
he churned and churned and churned
and turned the cream to butter —
then hopped out, alive
and kicking.

Ian Serraillier

99

Transformations

It was a tale of spells I told;
'Is it true?' she said,
'Is it true?'
The child's eyes grew round with every telling;
And still she asked,
'Is it true?'
'Is it really true?'
'Can the pumpkin turn into a coach?'
'Can the frog become a prince?'

Then I thought of transformations
I have seen
Or heard tell of;
Of boys become men,
And men grown old:
Of new blossom, falling leaf,
And, again, new blossom,
Year to year.
'Yes,' I said, 'it is true.'
'In its way,
 It is true.'

John Cunliffe

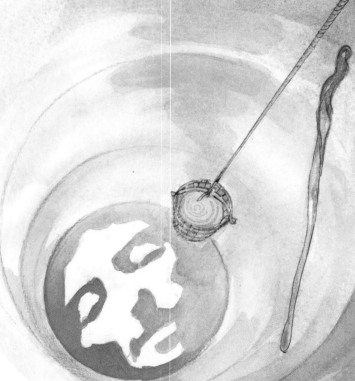

The Woman of Water

There once was a woman of water
Refused a Wizard her hand.
So he took the tears of a statue
And the weight from a grain of sand
And he squeezed the sap from a comet
And the height from a cypress tree
And he drained the dark from midnight
And he charmed the brains from a bee
And he soured the mixture with thunder
And stirred it with ice from hell
And the woman of water drank it down
And she changed into a well.

There once was a woman of water
Who was changed into a well
And the well smiled up at the Wizard
And down down that old Wizard fell. . .

Adrian Mitchell

Water and Ice

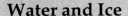

Water, you turn into ice
and put manacles on yourself.
You are your own prisoner.

You who used to sing and hum
are still as the grave,
white as a tombstone.

I slide all over you
where once I could drown.
Now I feel easy

on your white ballroom floor
just like a dancer
making figures of eight.

When spring comes you will die
ice of the winter.
Your handcuffs will fade.

There will first of all be stars,
then a brown sludge,
then a huge, lion-like roar.

Iain Crichton Smith

Solid Water

How can it be
That ducks walk
On the lake?
Why have the puddles
Turned to glass?
What magic makes
The stream stand still?
Why won't the upturned bucket spill?
Why do I slide each step I take?
What can forge spears
From blades of grass?
There's a fire inside me
That makes each breath
Like smoke unfold.
The slippery swords
From the window sill
Melt in the heat of my fire
But burn my hand with cold.

Julie Holder

103

Fireworks

First there was the eager looking forward
As hoarded in the safety of biscuit tins
Those bright cardboard buds awaited their flowering.
When, at a match's touch came
The sudden and immediate joy of crackle and colour
And the rocket's fiery upward rush.
Then last, the early morning's pleasure
(The night's smoke still on the air)
Of searching for the spent cases,
The charred and burnt out husks,
The scorched rocket dew-drenched on its stick.
Something that might actually have soared
As high as our hopes.

John Cotton

From Carnival To Cabbages and Rain

The narrow streets
Are smiles wide
Carnival has come to town.
Granny has a rose in her teeth
The baby wears a crown.
Everyone has come outside
To follow pied piper bands
Wearing dressing up clothes
Dancing hand in hands.
Hearts and blood
Beat to the drum.
Children free balloons —
'I gave mine to the sun'
A child cries.
Strangers are greeted as friends
Under blue skies.
The streets vibrate
Deep into the night
And rock from end to ends.
Children sleep on parents' shoulders
Late and light
Weaving Carnival into dreams
Round rainbow bends.

They shop for cabbages today
In narrow streets
Polite and grey.
Glitter shines
Down in the drain
And people say
'Now it can rain.'

Julie Holder

The Stone Owl

For Jo who found an owl in a stone

The stone owl waited
on the edge of the stream:
ten million years
he slept his owl dream.

He waited through water
and fire and snow
for the seasons to shape
the owl in his soul.

Then the small child came
to play in the stream
and reached for the stone
to throw — but his feel

warmed the stone owl awake
and the small child cried:
There's an owl in this stone
He's alive! He's alive!

Now the stone owl stands
looking out at the moon
from the high attic window
of the small child's room.

He has worn-away eyes
and a chipped-away beak
and heavy humped wings
and sharp flinty feet.

There he flies in his dreams
beyond any known bird
to a time long ago
when stones ruled the world.

He flies over oceans
that have dried up and gone,
over deserts where whales
once played in the sun.

He flies till he drops . . .
one more stone by a lake,
waiting for someone
to wish him awake.

Peter Bland

Breakfast to Dinner

At break of day
Roaring,
The lion claimed the kill.
Snarling
To keep the others at bay
He ate his fill.
Yawning,
He lay in the sun to sleep
And was still.

At noon
He does not wake to drink,
There is dust in his eyes
And he does not blink.
Flies walk
The furrows of his brow.
See the vultures wheeling —
He is their meat now.

Julie Holder

Elephant

I saw a picture
of an elephant
in the colour supplement
last Sunday.
It lay in folds
of grey across
two pages
Dying.

The wrinkled skin
which kept it in
has no more
worth.
But its tusks will play
Symphonies.

Karenza Storey

The Circus

A cloud of tawny dust rises
And covers the wide space between
The shiny river and the humble
White faces of the little houses.

The dust is so dry the lightest
Breeze will lift it, send it searching
Into your eyes and mouth and hair.
The breeze as suddenly comes to rest

As it sprang up, lets the dust settle,
And reveals in the centre there,
Bright as the sun and the blue sky
The caravans in a circle.

A man, black moustache smiling, puts down
A plate to feed the puppy tied
To the wheel. The puppy's tail wags
His whole joyous body. Swarms

Of brown lively children roam about.
Goats and barking dogs wander to and fro.
Women in gold and blue and scarlet
And green prepare the meal. Their hair

Like shining water sways and swirls.
The patient donkeys wander, trying
To find some weeds to graze upon.
Near one caravan some village children

Stand silent. The cage has black
Thick bars. It is so small the bear
Inside has hardly room to turn.
The fur has gone from his flanks and back.

Like some shabby clock, almost run down,
He shifts his sad weight from paw to paw,
While near him, utterly unaware,
Happy in the sun the travellers feed.

Albert Rowe

The Man on the Flying Trapeze

Sporting and capering high in the breeze,
cavorting about from trapeze to trapeze
is an aerial acrobat, slim as a ribbon,
as daring and free as a tree-swinging gibbon.

He hangs by his fingers, his toes and his knees,
he dangles and dips with astonishing ease,
then springs into space as though racing on wings,
gliding between his precarious swings.

He cheerfully executes perilous plunges,
dangerous dives, unforgettable lunges,
delicate scoops and spectacular swoops,
breathtaking triple flips, hazardous loops.

Then this midair magician with nerves made of steel
somersaults, catches and hangs by one heel.
As the audience roars for the king of trapezes
he takes out his handkerchief, waves it . . . and sneezes.

Balanced above us, the high wire king
skips with a swivel, a sway and a swing.
He dances, he prances, he leaps through the air,
then hangs by his teeth while he's combing his hair.
He seems not to notice the perilous height
as he stands on his left hand and waves with his right.

Jack Prelutsky

113

The Clowns

the clowns are not really happy
the clowns are not really sad
they just make mistakes
like me and you
maybe they didn't mean to be clowns
it was all just a big mistake

it's their mistakes that make us laugh
mistakes don't really make us happy
mistakes don't really make us sad
we laugh because we know
we make the same mistakes too

we're laughing at our own mistakes
we're laughing while we sit and watch
someone else making our mistakes instead

we're laughing because we dream in bed
one night we'll make the same mistake
and wake up in the morning to find
we've all turned into clowns

Dave Ward

Two In One

He peeled off his eyebrows,
Pulled off his nose,
Unlaced yellow boots
With extravagant toes,
Took off his top hat,
Green hair
And patched clothes,
He wiped the smile from his face
And
Where a clown had just stood —
A little old man
Appeared in his place.

Julie Holder

Family Photo

I can't believe it's my Dad
Standing there
His hair all golden
His bottom bare·
Two little dimples in his face

When now, all fat and bald
With straggly eyebrows
Which he just won't cut,
He's more like something
From another race.

John Kitching

An Old Snapshot

Who is this child, alone
in a waste of sand and sea?
Can hers be the flesh and bone
that scaffold me?

In that far summer I know
she saw the horizon clear,
and the sun go down below
as I do, here,

And she felt through her fingers fall
the drifting silt of time
who exists now — if at all —
in tearless rhyme;

And I ask myself, as I stare
at the photo of sand and sea:
is it I who am captured there,
or she here, in me?

Jean Kenward

Change Yourself

When he was very young,
the boy wanted to change
the entire world: but
he soon found he could not.

When he was grown up,
he tried to change those
around him — relatives and friends:
but soon found he could not.

When he was very old, he saw
how foolish he had been:
he at last began to change himself,
and in the end succeeded.

James Kirkup

Somewhere in the Sky

Somewhere,
In the sky,
There's a door painted blue
With a big brass knocker seven feet high.
If you can find it,
Knock, and go through —
That is, if you dare.
You'll see behind it
The secrets of the universe piled on a chair
Like a tangle of wool.
A voice will say
'You have seven centuries in which to unwind it.
But whatever
You do,
You must never,
Ever,
Lose your temper and pull.'

Leo Aylen

Kaleidoscope

Inside, this tube is
dark as a telescope.
But at one end, my eye
sees light at the other —
not stars or comets
but a rose window,
bits and pieces of
tinted jigsaw glass,
geometric, yet original,
agile arabesque, ever
changing, ever new,
organic as the pattern
in fresh sliced fruit.

My hand gives the tube
a gentle shake. The rose
window, with a tinkling
rustle, collapses on itself
yet is not broken, for
its jewelled segments
re-arrange themselves into
another sharp design, quite
different, but just as good —
a wheel of chance
and choice, transparency
of dreams, pressed eyelids —
stained glass of eternity.

James Kirkup

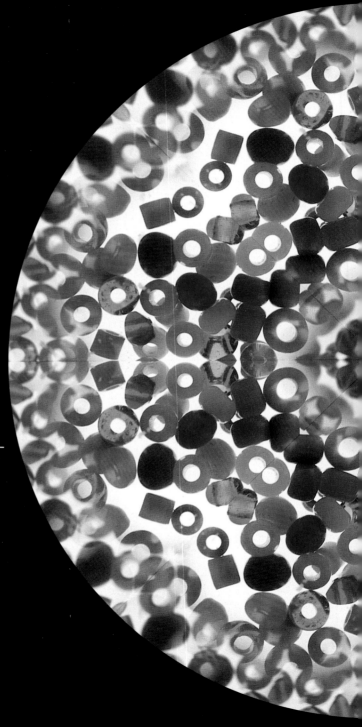

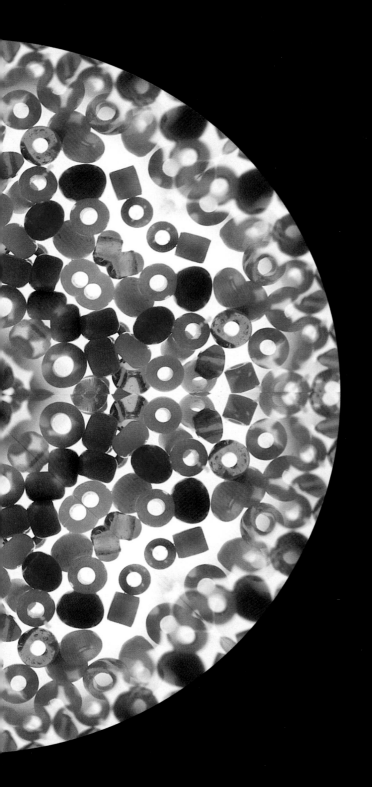

The Last Magician

A hundred miles beyond Beyond,
  Nearer than Here, further than Far,
The last magician breaks his wand
  And shapes the splinters to a star.

Then, leaning in the door of death,
  With his last art he forms a spark,
Breathes on it with dying breath
  And leaves it flaming in the dark.

A planet circles round that star.
  There Life evolves, and men who learn
The secretest of things, and are
  Master magicians in their turn.

A hundred ages pass. The sun
  Begins to die. The Wizard's skill
Makes hideous weapons. One by one
  They hunt and slay each other, till,

A hundred miles beyond Beyond,
  Nearer than Here, further than Far,
The last magician breaks his wand
  And shapes the splinters to a star.

Peter Dickinson

122

Index of first lines

Acknowledgements

We are grateful for permission to include the following copyright material in our anthology:

Moira Andrew: 'Autumn Sorrow' and 'Calendar of Cloud', both © 1986 Moira Andrew. Published in *A Calendar of Poems*, ed. Wes Magee (Bell & Hyman, 1986). Reprinted by permission of the author. Peter Bland: 'The Stone Owl'. Reprinted by permission of the author. First published in the New Zealand *School Journal*. N.M. Bodecker: 'Mosquito' from *Snowman Sniffles & Other Verses*. Copyright © 1983 N.M. Bodecker. Reprinted by permission of Faber & Faber Ltd., and Margaret K. McElderry Books, an imprint of Macmillan Publishing Company. Alan Brownjohn: 'Explorer', © Alan Brownjohn. Reprinted by permission of the author. Charles Causley: 'Green Man in the Garden' from *Collected Poems* (Macmillan). Reprinted by permission of David Higham Associates Ltd. Frank Collymore: 'The Spider'. Copyright Frank Collymore. Peter Dickinson: 'The Last Magician' from *Hundreds and Hundreds*. Reprinted by permission of the author. Max Fatchen: 'Night Walk' from *Songs For My Dog and Other People* (Kestrel Books, 1980), copyright © 1980 by Max Fatchen. Reprinted by permission of Penguin Books Ltd., and John Johnson Ltd. Robert Fisher: 'Uncle Fred', first published in *Ghosts Galore* (Faber, 1983), © Robert Fisher 1983. Reprinted by permission of the author. J.W. Hackett: 'A bitter morning: . . .' reprinted from *The Zen Haiku and Other Zen Poems of J.W. Hackett*, (Tokyo: Japan Publications, Inc., 1983). Distributed in the British Isles by International Book Distributors Ltd., 66 Wood Lane End, Hemel Hempstead, Herts. HP2 4RG. By permission of the author. Julie Holder: 'From Carnival to Cabbages and Rain' and 'Breakfast to Dinner'. Reprinted by permission of the author. Ted Hughes: 'The Snow-Shoe Hare' from *Under The North Star*. Copyright © 1981 Ted Hughes. Reprinted by permission of Faber & Faber Limited, and Viking Penguin Inc. Shake Keane: 'Once the Wind . . .' from *Wheel Around the World* ed. Chris Searle. Used by permission of Macdonald. Jean Kenward: 'An Old Snapshot', © 1966 Jean Kenward. Reprinted by permission of the author. J. Patrick Lewis: 'A Tomcat Is', © 1986 J. Patrick Lewis. Reprinted by permission of the author. Adrian Mitchell: 'The Woman of Water' from *Nothingmas Day*. Reprinted by permission of Allison & Busby. Ogden Nash: 'The Ostrich' from *I Wouldn't Have Missed It*, published in the United States in *Verses from 1929 On* Copyright © 1956 by Ogden Nash. First appeared in *The New Yorker*. Reprinted by permission of Andre Deutsch Ltd., and Little Brown & Company. Judith Nicholls: 'Frogspawn' from *Magic Mirror*. Reprinted by permission of Faber & Faber Ltd. Jack Prelutsky: 'The Invisible Beast' and 'The Darkling Elves' from *The Headless Horseman Rides Tonight*. Reprinted by permission of A & C Black (Publishers) Ltd; 'The Man on the Flying Trapeze' from *Circus!* Text Copyright © 1974 by Jack Prelutsky. Michael Rosen: 'Look Said the Boy' from *Wouldn't You Like to Know* Reprinted by permission of André Deutsch Ltd. Karenza Storey: 'I Saw a Picture' from *The Elephant Book* ed. Dennis Pepper (OUP, 1983). Reprinted by permission of the author. Michael Tanner: 'Magpie in the Snow', reprinted from *The Beaver Book of Animal Verse*. Louis Untermeyer: 'Adventure' extracted from "Jade Butterflies" from *Roast Leviathan* copyright 1923 by Harcourt Brace Jovanovich, Inc. Reprinted by permission of the publisher. Kit Wright: 'Whisper Whisper' and Lies' from *Rabbiting On*. Reprinted by permission of the publisher, Fontana/Collins.

The following poems are reprinted for the first time in this anthology and appear by permission of the author unless otherwise stated.

Leo Aylen: 'Somewhere in the sky', © 1987 Leo Aylen. Valerie Bloom: 'New Baby' and 'Chrismus a Come', both © 1987 Valerie Bloom. John Bond: 'Wolf' and 'The Chimpanzee', both © 1987 John Bond. Jacqueline Brown: 'Travelling Child' and 'Telephone Wires', both © 1987 Jacqueline Brown. Richard Burns: 'Dandelion', © 1987 Richard Burns. Dave Calder: 'to gain power over a balloon', © 1987 Dave Calder. Stanley Cook: 'Using a telescope', 'The Balloonists', 'Winter Journey' and 'Demolition worker', all © 1987 Stanley Cook. Pie Corbett: 'Owl', © 1987 Pie Corbett. John Cotton: 'Fireworks', © 1987 John Cotton. John Cunliffe: 'Transformation', © 1987 John Cunliffe. John C. Desmond: 'Lost in a Crowd', © 1987 John Desmond. Berlie Doherty: 'Circus School', and 'Frogspawn, I Love It', both © 1987 Berlie Doherty. Max Fatchen: 'So Clever, So Help Me!' and 'What a Calamity!' all © 1987 Max Fatchen and reprinted by permission of John Johnson Ltd. Eric Finney: 'Back to school blues', 'Simple Seasons', 'A Final Appointment', 'Wilkins' Drop', 'Wilkins' Luck', 'Whoppers',